Honest Reality

ENVIRONMENTALISM IS AN ANTISCIENCE CULT

David Noonan

ISBN: 0692677429
ISBN 13: 9780692677421

Introduction

This book does not question the passion or sincerity of the average environmentalist, but it does question their knowledge of reality. A lot of well-meaning people joined the environmentalist movement thinking that they were going to improve the world. They were wrong. Instead, they were lied to and used to empower and enrich the criminal elite. Although most environmentalists have been minor players in this massive swindle, their support of the con artist leaders has made the average environmentalist an unknowing enabler in the biggest and most expensive antiscience fraud in history. Our adventure into the real world begins now.

Contents

Part One

The Money Shots

The sun's mass is 332,946 times greater than that of the earth. The surface of the sun is upward of ten thousand degrees Fahrenheit of blazing heat. It also has occasional flash points of even greater heat called "solar flares" and "sunspots."

The earth's inner core is made of molten rock, which is also around ten thousand degrees Fahrenheit. Like the sun, the earth, too, has something similar to flash points, called "volcanoes," and it also has fault lines, which release heat on its surface and also under its oceans.

In between those two raging furnaces of fluctuating heat is the earth's outer crust. The crust is a system of floating and shifting tectonic plates that supports all of earth's plants, animals, and human beings. In spite of this constantly changing real world, where humanity lives between two raging furnaces, political hacks and their witch doctors claim human beings are changing the earth's climate. Manmade global warming is the most outrageous antiscience fraud in the history of the world. The claim of manmade global warming has become a worldwide criminal enterprise.

Today's environmentalist leaders act like domestic terrorists. They constantly threaten life-or-death destruction unless they are paid vast sums of money. They are extortionists who sell indulgences. They enrich themselves by selling scary lies. Like true criminals, they conceal the truth and provoke false panic so they can prosper as a result of the suffering of others. They use "problems," which they create, to establish tyranny and monopoly. They profit off of those lies with fake solutions, which are also lies.

Ponzi schemes start with a threat of loss—of opportunity, property, or of life itself. The supposed answer to stop that loss is to give con artists money and control. These criminals do not add value to anything, but most still claim to be beneficial to their victims.

Among the murderous is an old joke: that the death of one or two humans is a tragedy, while the death of millions is just a statistic. The environmentalist movement is built upon the corpses of millions of humans who were killed by the environmentalist ban on DDT. This ban was the start of worldwide suffering because of the environmentalists' liar laws.

In 1948, a Nobel Prize was given to Paul Hermann Müller, who discovered the use of DDT as an insecticide. Besides being cheap and easy to make, DDT was very successful for this purpose. In the late 1940s, DDT eliminated malaria in the United States and Europe. In Africa and Asia, it saved millions of human beings from being tortured and killed by nature. Some called DDT a "miracle weapon" against malaria.

In 1962, a book titled *Silent Spring* by Rachel Carson claimed and insinuated a lot of things about the supposed dangers of DDT to wildlife and humans. This book is the foundation of today's environmental movement.

Because of the claims against DDT, the government decided to investigate and sort out the truth. Besides being a very effective product against insects and the diseases they carry, DDT was found not to be very biologically active. In fact, the US National Toxicology Program classifies DDT as "moderately toxic."

Government scientists concluded that DDT, when used properly, poses no imminent threat to wildlife or humans. Despite this finding, the lead government bureaucrat at that time declared a ban on DDT, basing the decision on political correctness instead of the truth. It was a liar law. The ban started in 1972, and death rates shot up in the third world because other nations followed the United States' lead.

In 2006, the US government quietly resumed funding the international use of DDT to help stop the massive deaths each year. In 2007, an official of the National Institutes of Health stated that "The Ban on DDT may have killed twenty million children."

Popular author Michael Crichton, a stickler on accuracy, wrote in his book *State of Fear*, "Banning DDT killed more people than Hitler." In 2008, the World Health Organization (WHO) estimated that well over two hundred million people get malaria each year and, of those cases, more than eight hundred thousand die each year—most of them small children. WHO has what is called "vector control" to use DDT. It is just politically incorrect to use DDT.

The actual number of how many people died as a result of the environmentalists' ban on DDT will never be known, but it remains in the millions. In the anticivilization, the criminal class gains power and profit according to the number of people it hurts. The environmentalist movement gave birth to itself by killing millions of innocents. This was all done under a fascist concept called "political correctness," through which the mass killing of millions was hidden from the public.

A building is no better than the foundation it is built upon, and environmentalism is built upon lies and death. The big lie is the very foundation of today's environmentalism. The claim that DDT was banned to save human life is a criminal lie.

The environmentalists' climate swindle is a multistory house of corruption. Its foundation is pure evil, and its first floor carries on the con artist's tradition of lying. The fatal flaw of all master parasites is the belief that "might makes right." But if muscle counted more than intellect, dinosaurs would still dominate the earth. That is the terrible secret the consensus champions try to hide from the public. When confronting honest reality, numbers mean nothing. Here again the con artists try to sell what does not exist—consensus science.

"Consensus" is a political term for "mob rule." The majority can be right...or they can be wrong. When they are right, good things can happen. But when they are wrong, true dictatorship will rule. Consensus is an external authority the weak-minded especially embrace. Consensus gives the lazy an excuse not to think. If something goes wrong in the future, the nonthinkers can blame others for being wrong. It does not take brain power to join a mob.

Consensus means there will be no new innovations or new ideas. The truly new is not allowed. Consensus says so. Consensus is when everything is frozen in time. It is what the ruling class wants—a permanent status quo.

Consensus is for politics. People who build new things, create, and discover are not herd animals. Real scientists know the debate is never over, that there is always something new to learn. Those who claim the debate is settled are trying to use censorship to hide something.

Most environmentalists believe in political correctness, and that means they act like fascists. No one is allowed to disagree with them or have an opinion that is different from theirs. They will try to smear and slander opponents as being deniers, likening the dynamic to those who deny the Holocaust. This is very ironic since environmentalists themselves are among the biggest deniers of reality.

> ***There is no such thing as consensus science. If it's consensus, it isn't science. If it's science, it isn't consensus. Period.***
>
> **—Michael Crichton**

Real science is based upon skepticism. The environmentalists' claim that only consensus science matters is a deliberate lie. The environmentalists like to quote a couple of studies that supposedly prove that almost 100 percent of scientists believe manmade global warming is real. The problem is that each of those so-called studies has turned out to be falsified data.

In an online survey of only two questions, over ten thousand scientists were contacted. Just over three thousand responded, but only seventy-seven responses were chosen to pull as quotes. Of those, seventy-five answered "yes" that humans were changing the climate. Seventy-five divided by seventy-seven is 97.4 percent. The environmentalists' claim that 97 percent to 98 percent of scientists agree that humans are changing the earth's climate is a criminal lie.[1] The other 97 percent claim was also based upon falsified accounting.

The earth is continually reshaping itself. It always has and always will, and nature is always changing the climate. During the last glacial episode, the "Wisconsin glaciation," the sea level was low enough for people and animals to cross the Bering Land Bridge from Russia to Alaska.

The world's climate has changed topography so much over the past millions of years that there is a salt mine eleven hundred feet below Detroit, Michigan. Under the Arctic Sea ice are vast pools of oil, which the environmentalists claim is made from dead plants and animals. But man did not cause that drastic change in climate from glacial freezing

---

1 Watts, Anthony. "About That Overwhelming 97–98 Percent Number of Scientists That Say There Is a Climate Consensus." July 18, 2012. Wattsupwiththat.com.

to blistering heat, over and over again. Climate change is a natural reality.

The climate has always been changing and always will change, because it is what nature does. The environmentalists are the only ones who don't believe in nature's climate change. Like ancient rain dancers, these con artists tell everyone they can change the weather if people will just pay them. In their world of make believe, manmade climate change is real and is going to melt the glaciers and drown everyone who lives in lowland areas.

Myth: Melting Glaciers Will Flood the Earth

The claims that the glaciers are going to melt and flood the earth are hysterical. Nobody told the backward believers that the last glacial system melted over ten thousand years ago and that is why over 70 percent of the earth is already underwater. All that is left of the intercontinental ice sheets are remnants of what once was. Most of the ice left from the last Ice Age is in Antarctica.

Sixty thousand years ago, in the early Wisconsin glaciation, New York, Cincinnati, Saint Louis, Chicago, the Twin Cities, and points northward all through Canada were under a sheet of ice a mile or more thick. That glacial period formed the Great Lakes and filled the oceans with its melting.

Earth's tectonic plates are like rafts floating on water, but instead these plates float on molten rock. As the heavy glaciers melt, there is less pressure on the earth's crust where the glacier was located. The crust gradually rises because of the lighter pressure on it. Conversely, the greater weight of a big body of water forces the earth's crust downward. Earth's tectonic plates constantly shift because of pressure from the surface and below the surface. This affects sea level.

Only land-based ice, which is miniscule to what once was, can possibly add to the sea level. When floating ice melts, there is no change in water level since its liquid volume is less than its solid volume. Contrary to what the fear-mongers say, if the Arctic ice pack were to melt—or a drinking glass completely full of water and floating ice melted—there would be no flooding or overflow.

Test it with a plastic water bottle:

- Fill the bottle half full, close it with its lid, mark the water level, and freeze it overnight.
- The next day, you will see that the ice line is above the marked water line.
- Leave the closed bottle out to thaw, and the ice line will go back to the original water line.

When water freezes, its volume increases. When ice melts, its volume decreases. When floating ice melts, the water level stays the same. Those who claim floating ice that melts will raise the water level are either ignorant or they are liars.

To understand the next lie, some background is needed. Winter is caused by the earth's axis tilting away from the sun and summer is caused by the axis tilting toward the sun. The average Arctic winter temperature is around minus forty degrees Fahrenheit, and the average Antarctic winter is around minus seventy-six degrees Fahrenheit. The reason the South Pole winters are colder is because at that time of the year the earth is 3.29 percent farther from the sun and its heat. The South Pole is also colder because the Antarctic is mostly land based and is higher in the cooler air.

Besides the floating-ice lie, some environmentalists also claim manmade global warming will raise the earth's temperature two or three degrees Fahrenheit, melt the polar ice caps, and flood the world. A two or three-degree increase will not make the ice caps disappear. Ice cannot melt until the temperature gets over thirty-two degrees Fahrenheit. The claim that a two- or three-degree increase will make the polar ice caps vanish is not true.

Myth: Carbon Dioxide is Out to Get You

The next story in this house of cards involves carbon dioxide (CO2). In true form of the upside-down world, some environmentalists claim nature's greenest and most life-giving gas is poison and a toxic pollution. But in the real world, carbon dioxide is a natural and necessary part of nature. Without carbon dioxide, planet earth would die.

Carbon is the building block of life. Earth is a carbon-based planet. Terra (land) is made from carbon, and all flora (plants) and fauna (animals) are carbon life forms. Carbon dioxide is created by the decomposition of carbon, and this continuously happens throughout the earth.

Carbon dioxide is released from the decay of organic matter and the breakdown of other carbon-based materials. Here again some people claim carbon dioxide produces warming, but in the real world it is the opposite. It takes warming to produce carbon dioxide. Heat is necessary to release carbon dioxide from carbon by breaking it down. Carbon dioxide does not cause heat—it is the by-product of heated carbon. Carbon dioxide does not control the earth's temperature, as the sun is the main influence over earth's surface temperature.

After standing in hot summer sunlight and then moving into a nice shaded area, common sense tells most people the sun is the main heat source for the surface of earth.

What many environmentalist leaders will never tell the public is that nature puts twenty-four units of carbon dioxide into the atmosphere for every single unit man puts into the air. Schools used to teach about the power and necessity of photosynthesis in the cycle of life. People inhale oxygen and exhale carbon dioxide. Plants consume carbon dioxide and release oxygen. Carbon dioxide is plant food, and the plants cannot live without it.

Environmentalists leave out the fact that nature itself puts 96 percent of all carbon dioxide into the atmosphere while humans put only 4 percent of the trace gas into the atmosphere.

Carbon dioxide is emitted naturally; the earth's own geologic actions place most carbon dioxide in the atmosphere.

Carbon dioxide has a life cycle very similar to that of water. Neither is poisonous but both can kill in the same manner—through suffocation. Carbon dioxide is nontoxic, odorless, and colorless.

All living creatures exhale carbon dioxide. It comes from the digestion of food. Carbohydrate is carbon. The normal level of carbon dioxide humans exhale into the air is about 4 to 5 percent. This means human beings can live with up to fifty thousand parts per million (ppm) of carbon dioxide in their lungs. People exhale over forty thousand ppm of carbon dioxide all day long every day without any ill effects.

Submariners don't pay attention to carbon dioxide levels unless they start to go over eight thousand ppm. The human body is designed to automatically exhale when carbon dioxide exceeds 5 percent—otherwise the body will start to suffocate. Those with sleep apnea experience this. When the body stops breathing, the brain wakes it up so it will not drown in carbon dioxide. Carbon dioxide is about half again heavier than air. It can settle in caves, silos, and quiet low spots.

Environmentalists who claim anything over four hundred ppm of carbon dioxide in the atmosphere will melt and destroy the earth are citing voodoo science. With each breath exhaled they are in fact emitting over one hundred times the amount of their imaginary disaster.

All vegetation needs carbon dioxide to live, even water plants. Greenhouse owners inject carbon dioxide into their hothouses to make their plants grow bigger and better. During the Jurassic period, plants and animals were huge, numerous, and flourishing. Atmospheric carbon dioxide was over two thousand ppm back then. All that extra carbon dioxide allowed things to grow bigger than they can today. Even when carbon dioxide is at eight thousand ppm or more, it does not threaten human life. Carbon dioxide is even used in carbonated beverages and is listed on the back of soda cans as carbonated water.

Almost half of a dried plant is carbon, and most of that comes from carbon dioxide floating in the air. Plants can grow out of cracks in rock cliffs because their food is airborne carbon dioxide. The only things a seed needs to grow are temperate weather, water, sunlight, and carbon dioxide. Carbon dioxide is floating fertilizer. To declare war against it is to declare war against earth itself. Carbon dioxide is as natural as rain. And yet some claim that anything over four hundred ppm will destroy the earth and mankind. These claims are simply untrue.

Myth: Global Greenhouse Theory is a Real Threat

Today's manmade global warming hoax is modern man's tulip bubble. The global greenhouse theory holds and gives shape to the environmentalists' house of cards. It is the framework that holds the climate swindlers' stack of lies together. In the real world, this theory has the tensile strength of a single sheet of toilet paper—used. It is the biggest lie of all.

British scientist John Tyndall is officially credited by historians for starting the global greenhouse theory in 1859. This was before the Civil War, which was fought with primitive black powder firearms. At the turn of the century, the global greenhouse theory was discredited. It was brought back in the '50s and '60s.

Global greenhouse theory states that trace gases in the atmosphere will trap heat coming up from the earth, reradiate it back to earth, and elevate the earth's surface temperature. The more so-called heat-trapping gas in the atmosphere, the more heat is reradiated back to earth. There are several fatal flaws in this corrupt theory.

The following four flaws, by themselves, do enough to expose the fraud of the global greenhouse theory:

The claims that trapped heat in the atmosphere flows back to earth is an impossibility, as it violates the law of thermodynamics. Lesser heat cannot flow into greater heat. Like water flowing downhill, heat can move only from a higher to lesser temperature. A candle will not radiate its heat into the heat of a blowtorch and make the blowtorch hotter, any more than a creek emptying into a river will flow upstream. If heat could be amplified according to the global greenhouse theory, the world would have a perpetual energy source that would forever solve the issue of energy supply.

Another fatal flaw is the claim that atmospheric gases trap heat. Gas cannot trap heat. Gases are molecules, liquids are molecules, and solids are molecules. Each molecule has its own density. Heat flows through the molecules according to the molecules' resistance. The term "absorbing heat" is a misnomer. Molecules do not absorb heat, and they do not trap heat. They only slow down the heat's passing based on their own molecular density. Heat passes very quickly through molecules floating in the atmosphere. With solids and liquids, the passing of heat is slower. This slowdown can cause the warming of the molecule, but when the heat source is cut off, the molecule cools down to what it was before. A metal frying pan on a hot burner can get good and hot, but when the heat source is turned off, the metal frying pan soon goes back to room temperature. A metal frying pan cannot trap heat and neither can gas. The same is true of earth. Nighttime proves the earth can cool off very quickly. The atmosphere's gases do not trap any heat. They only slightly slow down the passing of heat, like a breeze passing through a screen door.

Diagrams that portray global greenhouse theory present only part of the whole picture, and few people look at the missing piece, which is that half the time half the world is in the dark. There is no incoming heat from the sun to warm the dark side of the earth. When this happens nighttime temperatures can be several degrees cooler than daytime temperatures. This proves two important things. *First*, the sun is the primary heat source for the earth's surface. *Second*, when that heat source is cut off, the surface temperature drops. More gases in the atmosphere, like water vapor, slightly slow down heat escaping from the earth. But it is also true that increased gases slow down and reflect incoming heat from the sun. There is no "global greenhouse" keeping the earth warm. If the earth suddenly stopped rotating, the dark side of the planet would turn into an ice ball and the sunny side would become a baked desert. Luckily for all living things, the earth rotates every twenty-four hours to regulate its temperature.

Every day around the world, temperatures rise and fall several degrees from day to night, and this fluctuation has caused no disaster. Yet environmentalists claim a two-degree rise in temperature will cause the end of earth. The surface of the earth gets energy not only from the sun but also from its own core of molten magma. The energy man makes is infinitesimal compared to the energy the universe makes. A changing climate is way beyond mankind's control for the foreseeable future.

During a hot summer rainstorm, water vapor can be pushed up several thousand feet into the atmosphere and fall back to earth as frozen hail. If the heat-trapping greenhouse effect were true, this could never happen. According to the theory, the atmosphere would be too hot to freeze the vapor because of all the trapped heat in the atmosphere. In reality, the temperature at thirty-five thousand feet is around minus fifty-five degrees Fahrenheit. Mountains and their tops are like reverse thermometers. The higher up one travels on the mountain, the colder it gets.

So exactly where is the trapped heat? With today's instruments it should be easy to say "X marks the spot." And yet no one has shown where the trapped heat is located, and they never will. There is no trapped heat because it is a physical impossibility.

A multitude of variables affect the earth's surface temperature. The strongest factor is the sun, a ten thousand-degree furnace that is 332,946 times bigger than planet earth. The sun's energy usually varies slightly but the sun's magnetic field can alter the output of radiation hitting the earth. More cosmic rays help clouds develop, in turn reflecting more solar radiation away from the earth to help cool the planet. Fewer cosmic rays let in more radiation, heating things up. The sun's magnetic field controls the cosmic rays' intensity, which controls surface water vapor. Water vapor is a large percentage of atmospheric trace gases. The fewer clouds in the sky, the faster heat escapes from the earth at nighttime, causing bigger swings in temperature. Large temperature variations destroy plants and animals, turning earth into a desert. However, large amounts of clouds cool the earth and deposit more water on the land, causing a new ice age.

Earth has its own variable magnetic field, which can also affect solar winds. Earth's magnetic field helps deflect high-energy particles from the galaxy. This field protects the earth from the sun's excess energy, making the planet livable. There is no imaginary greenhouse.

The earth's core is ten thousand degrees of shifting molten rock, which can also affect its surface temperature.

Volcanoes, hot water springs, tectonic plate fault lines, and conduction all release varying heat to the earth's surface and under its oceans. Heat is drawn from the earth's center, where it is ten thousand degrees Fahrenheit, out into far space, where it is minus 460 degrees Fahrenheit. The law of thermodynamics says so. Hot travels to cold, and in between is humanity.

On top of the environmentalist lie factory is the penthouse where the leaders reside and plan their fraud against the public. They do not want the public to know there are snowstorms in the tropics. A warm ocean surrounds Hawaii, and yet snow commonly falls on the islands' highest mountain peaks. There are glaciers almost on top of the equator. Kilimanjaro in Tanzania and Mount Kenya in Kenya have glaciers on them. Under the right conditions, summer or winter, ice can fall from the sky anytime and anywhere on earth.

The swindlers do not want the public to learn that their climate models are provably wrong, or that the data put into them is also corrupt and falsified. Some government agencies have been caught adjusting and falsifying weather temperatures to get the effects they want. Liars of all stripes have doctored the data to get more taxpayer money and private grants to champion the fraud called "manmade global warming."

The entire manmade global warming story is an empty bag of make believe. There is nothing there. It is all lies.

The atmosphere is so cold that plane engines can produce vapor trails of ice crystals because internal combustion engines create water vapor in the exhaust. The earth's atmosphere is so cold that when jet planes fly through the water vapor in the atmosphere, the planes can leave trails of ice crystals floating in the sky. At high altitudes, airplane wings can ice up even in the middle of summer since the high altitudes are freezing.

This is why high mountain peaks have ice and snow on them year-round. If there was heat in the atmosphere, ice and snow would not be on the tops of the taller mountains all the time—the alleged trapped heat would melt them.

The problem with telling one lie is, to keep it believable, more lies have to be told to make the first lie seem real. That is the nature of Ponzi schemes. The lies keep on building until the fraud becomes unsustainable and it collapses from its own weight of absurdity. The climate swindle is a complete fraud run by professional liars. Each liar has the dictator's dream of global governance with themselves in charge, collecting the money.

Even the global greenhouse theory itself is a 150-year-old superstition. The climate swindle is a massive fraud used to cheat and harm people for the benefit of a handful of elites. The idea that environmentalism's main concern is to protect and enhance all human life is a grand ruse.

Environmentalist leaders have become betrayers. They betray science, and they betray people. They continually create false debts and lay those make-believe obligations at the feet of the public. Then they demand those debts be paid to them right away. “Action has to be taken right now!” they shout, desperate to get the money before their swindle is discovered.

When con artists have trouble selling their “overwhelmingly accepted” lie, they have to rebrand it. The ’70s “advancing global ice age” later became “global warming,” which then became “climate change.” Then it became “climate disruption,” and now it is back to “climate change,” a term that sells better. To keep their confidence game going, the swindlers conflate “climate change” as being the same thing as “manmade climate change.” This is the flimflammers’ trick of claiming their lie is of equal value to the truth.

In the real world, climate change is normal and natural. Manmade climate change does not exist.

In true liars’ form, the environmentalists now say manmade climate change is as real as gravity. Again, this clearly demonstrates corruption by using conflation to claim an unprovable effect is as real as a provable effect. Environmentalists demand that the world accept their fraud as something very valuable. Instead of honestly managing nature, the environmentalists destroy nature for political and financial profit. They have declared war, for example, on nature’s greenest, most life-giving gas. The only thing these con artists care about is getting a paycheck. They don’t care if it is a financial paycheck or a psychological paycheck, as long as they can make a profit.

Like all other criminals, environmentalist leaders steal money for a living. They survive by lying, cheating, and extorting real values from those who work for a living. Con artists take the important ideals of civilization, turn them upside-down, and twist them backward. The campaign against manmade global warming has nothing to do with protecting people or the planet. It has everything to do with robbing the public and enriching the criminal elite.

Not only is the environmentalist movement a house of cards built upon the dead bodies of millions of innocent people, but the leaders use the same demand other terrorists and robbers have used throughout history.

"Give us your money or you are going to die!"

Part Two

The Anticivilization

The criminal elite cannot allow the public to discover how their criminal world really works. Understanding how the ruling class manipulates its victims is the master key to destroying their sick world.

When navigating through life, having a blueprint of basic reality is a must. The world runs on philosophy. Even though it controls and determines everything, philosophy is not a complicated subject. There are only two real sides to all the "isms." They are *individualism* and *anti-individualism*. Societies based on individualism are civil, healthy, and prosperous. However, anti-individualism societies are not civil—they are criminal. They rot, fester, and self-destruct into hate-filled violence. As the numbers and power of the criminal increase, the standard of living for everyone else decreases. Those in such a parasitic society do not see people as human beings. The uncivilized think of their neighbors as mere livestock who need to be herded, sheared, and eaten.

The two philosophies are separated into two opposite worlds—civilization and anticivilization—each ruled by a government. In a civilized society, government is truly a servant of the people, but in the anticivilization, government is the master of its citizens. That is the difference between a free society and a tyranny.

The civilized simplify and clarify, while the uncivilized do the opposite. Criminals are liars and cheaters who love to take the simple truth and obscure it so most people cannot recognize it. They profit by selling lies to the public.

The civilized control, manage, and try to work with nature for the betterment of themselves and their neighbors. The uncivilized attack and exploit human beings to make their living. They profit by causing others to suffer.

When plunder becomes a way of life for a group of men living together in society, they create for themselves, in the course of time, a legal system that authorizes it and a moral code that glorifies it.

—Frederic Bastiat

The entire leadership of the anticivilization lives on crime, and they encourage followers to also do criminal acts. The uncivilized prove to the rest of the world that evil is not an accident. Evil is deliberate.

The anticivilization is distorted reality. It is a world that is upside-down and backward. Its residents are aggressive predators, who care only about themselves. They live in their own fantasy world, where they think they are the supreme rulers over everyone else.

The criminal class does not work for a living—they steal for a living. They feel it is easier to destroy than to create. Criminals are problem makers, while the civilized are problem solvers. All crime comes from acting uncivilized.

Within the criminal class there are two subdivisions: common street thugs, who specialize in physical violence, and white-collar thieves, who specialize in financial and intellectual crimes such as fraud and extortion. Street criminals operate under the cover of darkness, while white-collar criminals operate under the cover of fraud.

Common street criminals make no pretense about what they do for a living. They openly rob, rape, and murder their victims. As a result, they are constantly hunted down and put in jail. White-collar criminals are different. They hide their crimes by keeping their victims hopelessly ignorant of what is really going on. These criminals think they are safe because no one has publicly exposed their liars' code.

The criminal class that runs the anticivilization is an affiliation of con artists, master parasites, and other professional liars who consider themselves part of the ruling class. No matter what name is used to describe the criminal class, all the different names have the same unifying theme. They are all evil and antihuman. They can stay in power only as long the public allows them to get away with their treachery.

The anticivilization is a system of shakedown schemes. The goal of all its con artists is to steal their neighbors' property and life savings. To accomplish this, the criminals look for ways to give the appearance of legitimacy to their aggression against the innocent. These haters need excuses to rationalize and hide behind.

The pretense of "doing good" is the cover they seek. To survive and pull off their swindle, con artists pose as saviors and do-gooders. This is why liars are so hostile to the truth. It shatters their fantasies. Con artists can only survive by manipulating people into making bad choices for themselves. Next to their threat of violence, illusion is the only weapon they have. One of the con artists' main tricks is to get people to focus on the wrong things. Fraud is designed to *not* be understood. If people knew what was really going on, the con artists would be headed to prison.

In the anticivilization the greatest truth of mankind is extinguished forever. No one believes all people have equal rights and that they are under the rule of equal law. In the anticivilization, some people are more equal than others. To make this happen, the public is deceived into accepting liar laws and a two-tier justice system. The "chosen ones" rule and control those who are not chosen. Individual rights are destroyed and replaced with special privileges.

There is no rule of honest law in the anticivilization. Honest law is replaced with liar laws. These laws are the opposite of what is true and just. They are the foundation the anticivilization is built upon. Liar laws are criminal weapons used to institute mass slavery. They are used by the few to exploit the many. Liar laws are not designed to help the public but are made to protect the criminal class. They are used to funnel commerce into elitist-owned monopolies designed to create illicit profits for the criminal class. In the anticivilization, master parasites write the laws for themselves, have their hireling journalists publicly champion them, and then have their bought politicians pass them. The anticivilization is a monopoly of control by the criminal ruling class. Laws and regulations become so numerous and ridiculous that eventually very few will obey them. The dictatorship of the anticivilization will turn into full-blown anarchy and become a superhighway into chaos.

The collapse of the anticivilization is brought on by its own corruption and greed. The only way the criminal class can survive is through subsidies, graft, forced monopoly, and above all, government protection. When the master parasites become so greedy and ravenous that they consume most of honest government, the end is near. The snake has started to eat its own tail.

The criminal class calls truth a "conspiracy," but they themselves use conspiracy against the innocent. Ponzi schemes are built by murdering the truth in order to get away with harming others. The environmentalist movement is a perfect example of how swindlers manipulate followers. Up until now, some environmentalist leaders have been very good at telling big lies and getting away with it.

Political correctness, or thought control, is the only thing that can keep con artists in power. Political correctness is a hate movement designed to destroy the free choice of the individual. People haters use hate speech to intimidate and bully anyone who dares to disagree or question their holy views.

The anticivilization is one big concentration camp divided into units, called "Ponzi schemes," that are run by master thieves. The environmentalist climate swindle is just one of many schemes run by these con artists. All of these frauds are interwoven and run by members of the same club. They might have different names and descriptions, but behind their masks they are all the same monster. They are unlimited evil.

Until the public chooses to turn away from fraudulence, the public's living standard will continue to drop until there is nothing left for the criminals to steal or destroy. The public living standard collapses because people are being cheated out of their labor. People are tricked and swindled out of the real values they produce. Working people are tricked or forced to exchange their real values produced by real work for counterfeit values produced by liars. That is how the public is defrauded. The parasites steal chunks of their victims' lives until victims have no lives left to steal. The anticivilization is a society of madness, hate, and total destruction.

The only thing that keeps con artists in power is their ability to fool others into believing they are legitimate. Once con artists are exposed as frauds, their power is gone. A society held together by ignorance has more than failure in its future. The sum total of treachery in a Ponzi scheme will give the future amount of pain and disaster that will be leveled on everyone who believed in it. Only one thing can keep the anticivilization and its parasites up and running—mass ignorance of honest reality. The general public has to be kept completely unaware of how they are being used and abused by the criminal class.

Ponzi schemes are naturally attractive for the uncivilized. The big lie of a Ponzi scheme gives people haters the chance to appear proper and superior to everyone else. The haters have an illusionary cloak of self-righteousness, while espousing hatred for their neighbors. Instead of being beat up and run out of town, the poseurs are cheered on and given awards. For many, joining the swindle club allows them the opportunity of being cruel to others and getting away with it. This happens all under some pretense of serving the greater good. It is no overstatement that public ignorance is the lifeblood of the criminal ruling class.

Elites designed the anticivilization to be a rigged society plundering the rest of the world, using Ponzi schemes as the main weapons. Instead of focusing on making their own wealth, the master parasites obsess about stealing their neighbors' wealth. Initiating a Ponzi scheme is an act of war against the civilized. The schemes create false opportunities or fake problems and then offer fake solutions to that problem. The greatest trick the con artists perform is to appear as necessary saviors, but reality is the opposite. Those usurpers will pay dearly when the public learns the truth.

Ponzi schemes are not accidents. They do not just happen. Everything in a Ponzi scheme is deliberately designed to destroy the innocent and good. The most important factor swindlers look for is ignorance in specific areas. The easiest victims do not like to think for themselves. They obey external authority without serious question.

For a Ponzi scheme to work, people have to believe in it. Ponzi schemes run on confidence. When that confidence is publicly questioned, the scheme trembles. And when that belief system is exposed as a fraud-based confidence scam, the game is over.

To start a Ponzi scheme, swindlers tell a big lie to see if it sells to the public. If it does, they keep adding more lies to it to draw in more suckers. Lies come from false values and are used to create false debts and fake obligations. Lies are false claims against other people's assets.

Lies, backed by the force of fraud or violence, are the main currency of the anticivilization and all the various swindlers within it. The innocent are held responsible for the lies of the con artists. The swindlers are issuers of false debt and false obligations. Their victims are told they owe something to someone else, and sometimes the marks are even told they owe it to themselves. In the real world the swindlers' claims have no value at all. Everything is fake, like the climate swindle, and yet victims of the swindle are told they must pay real assets to the con artists. Criminals survive by exchanging fraud for real wealth. Con artists are merchants of false debts. The more fraud they can sell, the richer they become.

Usurpers live off of confused victims. To keep their victims unsettled and off balance, the schemers constantly lie to everyone around them. The Ponzi cultists use double talk and scare tactics to get what they want. Terrorism is the con artists' game. Fear and panic are the liars' tools, and their criminal lies are acts of deliberate terrorism. Sacrifice is the fear merchant's favorite demand. They threaten that, if they do not get their way, the world will crash into ruination. They claim that disaster will all be the fault of those who fail to submit to their demands. The producers are forced to subsidize the takers. But in the real world, the swindlers' fake claims are in fact a false tax upon an honest society. As the numbers and magnitude of false debts increase, the living standard of the producer society decreases. This is why white-collar criminals have to keep presenting a make-believe reality upon the rest of the world. To be successful, the criminal class is in a continuous aggression against their neighbors. They need massive distractions for their victims to focus on. Con artists know they are only as important and powerful as their victims allow them to be.

All con artists are counterfeiters. Counterfeiting is stealing life from others. Fake values are forcibly traded for real values. Honest, hardworking people, who put their lives into producing real values, are brutalized and robbed by the uncivilized. The criminal class survives by selling fraud to the public. When criminals are allowed to take over a society, that society becomes the anticivilization, where no lie is too big to tell. Soon the standard swindle becomes an institution, where con artists are allowed to literally sell toxic garbage to the public.

To succeed, the ruling class cheaters need their victims to follow the rules of their crooked games. To obfuscate reality, con artists act like magicians and use misdirection. The military calls misdirection a false flag. The enemy is tricked into believing make believe and then acts against its own best interests. False flags are what all swindles are made of, and the environmentalists' climate swindle is a big bundle of them.

False flags are an act of aggression through fraud.
False flags are used to whip up irrational hate and panicked obedience. They are used to frame the good and the innocent. Liars specialize in frame jobs. They will say anything to get their true believers to act against their own best interests and harm others. Throughout the politics of the uncivilized, lies are used to start aggression against the innocent. Every lie, every false witness and every fraud is a planned fake out.

In the anticivilization, control freaks are in charge and innocent human beings are the enemy. To destroy the individual, the ruling class keeps people separated with false differences and keeps them attacking and hating one another over issues that are fraudulently founded. This is done so the public cannot see what is really going on. Instead, the con artists' victims focus on effect and not the real cause of what is happening to them.

The control freak elites are a self-anointed master race who think they should own and control everything and everyone. The anticivilization is a monopoly of control, and its elitists are so delusional that some think they are divine. While the civilized use reason to acquire things they want, control freaks initiate the violence of fraud and intimidation to get what they want.

The usurpers have to continually work to suppress public awareness. These control freaks will do anything to gain and keep control over others. They will even lie by omission. They are quite good at telling half-truths designed to mislead. They must be believable, but when their trustworthiness is questioned and starts to erode, they begin to lose their power to swindle. This is why con artists demand absolute control over information. Without self-censorship and government censorship, Ponzi schemes fall apart. Reality becomes obvious.

The anticivilization is a society based upon fear of the real. Here it is common for people who tell the truth to be threatened with slander and physical harm, all to protect the big lie. But the fear mongers are the ones who are really afraid. They know that if their own criminality is publicly exposed, they are done for. They know all they have is a fragile illusion. For them, it is imperative that they control their image of righteousness. Once their prior claims are exposed as complete fraud, the con artists themselves become worthless and powerless.

When things become political, big mistakes will be made because bias is placed over reason. As an institution becomes more political, it becomes more corrupt. Environmentalism is a perfect example. It has become a political movement based upon voodoo science and hate. The environmentalists' climate swindle is a total fabrication used by usurpers worldwide to gain illicit money and power. Environmentalist hijackers have turned the conservation movement into a grand system of robbery. It has become a cover to hide behind and a tool to exploit neighbors. The hijackers took what was once good, needed, and beneficial to all and turned it into something exploitive and evil.

Environmentalism is repackaged tyranny wearing a different mask. Like all other dictatorships, it is morally and intellectually bankrupt. It is a bully culture, in which the truth is the opposite of what is stated.

Fraud is the paramount glue that holds the climate swindle together. Gross nonsense is the foundation, structure, and framework of the climate hoax. Like all other major Ponzi schemes in the anticivilization, the environmentalists' swindle is pure fraud from bottom to top.

Some climate swindlers like to claim that their criminal enterprise is not a conspiracy and that such a conspiracy would be impossible because it would involve so many people. They are liars. Nazism and communism were mass conspiracies, just as the environmentalists' climate swindle is a mass conspiracy. There are plenty of sick and twisted people who would love to get away with expressing their cruelty and hatred for their neighbors. Nazi true believers loaded human beings into cattle cars, marched them into make-believe shower stalls, and gassed them to death. The communist true believers herded nonbelievers into the insides of barges and then sank the barges to save bullets. Asian communists marched victims into the killing fields, bound them, and tightened plastic bags over their heads. And now environmentalists threaten nonbelievers with slander, censorship, imprisonment, and worse. (See bibliography).

Worldwide conspiracies, like climate fraud, are supported by thousands of benefactors. Each and every supporter imagines they will personally profit from the big lie. The true believers fail to realize that they are hurting themselves. Crimes committed under conspiracies are all done under the guise of the supposed greater good, economic justice, social justice, or some other load of statist hubris. The "do-gooders" are very careful not to talk about individual rights because their own argument would fall apart. In reality, mass conspiracies are about the many being enslaved, robbed, tortured, and destroyed for the benefit of the few.

The only way centralized planning can grow is by destroying the individual rights of the many and concentrating that power into the hands of those who consider themselves superior to others. The more innocent lives they harm and destroy, the richer and more powerful the usurpers become. Nazism, communism, and environmentalism are all based upon fraud and aggression. All three champion dictatorship over the individual. "Environmentalist" has become just another word for political hack who has a scheme to lie, cheat, and steal like the rest of the other criminals.

Anyone can own these liars when they understand the lie and its falsity. The environmentalist climate swindle is a porthole into the very soul of the anticivilization. The climate fraud is a crystal-clear example of how the criminal class robs and brutalizes its victims. It is all done with condescending smiles upon the master parasites' faces. The environmentalist leaders are the spirit of evil bent upon the betrayal of humanity. Wherever they go, problems develop. Crisis and panic increase. Just like the Nazis and communists, the environmentalists constantly lie, bully, and demand sacrifice from others, all the while pretending they themselves are saviors of society. All three hate movements are comprised of a stack of lies backed up with liar laws that try to institutionalize the elites' criminality.

The leaders of the Nazis, communists, and environmentalists each used their own brand of witchcraft science to cast an evil spell upon the public. By instilling mass hysteria, they were able to exploit humanity. Each swindle was very profitable for insiders and grafters. All these criminals had to do was murder the truth and keep it suppressed. "Credibility" is everything for the con artists.

Change for the Better Is Imminent

The world is starting to change for the better. More and more people are waking up to honest reality. They now know for sure that there is good and evil, not just gray. They know that, in both worlds, each person becomes his or her own product, the embodiment of what he or she sells to the public. People become their deeds.

The civilized world has a universal language called "honest reality." Honest reality is knowing how things really work. Life is all about values. Honest and real values enhance human life. There are different accents and dialects, but the language of the civilized is the same everywhere—that is, to speak the truth.

Fraud can only exist through the suppression of honest knowledge. In the anticivilization, telling the truth is an act of sedition. People know when they are living in the anticivilization. They fear asking honest questions, and they are attacked for telling the truth about the corruption around them. In the inverted world of the uncivilized, telling the truth is called hate.

The template used for all swindles follows the same pattern: Tell the big lie. Threaten people with that fraud. Collect money from the suckers. Repeat the process.

Honest reality is a game of real knowledge. When a person learns that life is a game of wit, it becomes much easier to defeat the witless who believe in fantasy and superstition. The witless live in an upside-down world of illusion and delusions. They will learn that the criminal ruling class betrays everyone who believes in it.

Part Three

Common Brainwashing

Education is not the learning of many facts, but the training of the mind to think.

—Albert Einstein

There have been advances in technology but few true advances in human thinking. This is because the ruling class tries to keep the masses in subservience to them. The only way the elite can stay in power is by teaching that they are indispensable. Besides using mass media obfuscations to stay in power, the usurpers also use an education system based upon memorization.

Thanks to the memorizer system, the general public is conditioned and can be trained to believe in the unbelievable. Students are not taught to listen to their own conscious or inner self. Instead they are trained to be and act as a common external consciousness. They are told not to feel their own thoughts, to ignore their own inner voice, and instead automatically accept external authority. The individual's perception is crippled along with his or her humanity. This process produces self-censorship and reinforces state censorship.

Memorization is not truly the primary source of a real education. Memorization is repetition, and repetition is used for brainwashing. Brainwashing keeps the status quo in power. Memorization is all about control by the control freaks. It is a learned system of self-deception, used for indoctrination.

The real purpose of a memorizer education is to instill ignorance. The student might be able to recite fact after fact but is unable to figure out which of the so-called facts are real or how they were arrived at. Students are trained to be oblivious to how and who is manipulating them. The sole duty of the memorizer is to become food and cannon fodder for the criminal elite. All usurpers know that, if they can control the education system, they can control the entire population. Dictatorship can only survive when the dictators do the teaching.

The memorizer system is an artificial world of imaginary education run by the status quo, not the honest and knowledgeable. Like the rest of anticivilization, it is a hub of monopoly and centralization run by elites for elites. Here students are taught that they are worthless without external authority guiding them and that they themselves are only a means to an end.

The memorizer system is designed to be educational oppression. It instills cognitive dissonance through which reality is what the ruling class says it is. Any questioning or skepticism about the official decrees of the state is forbidden and attacked as extremist. Absolute obedience to the criminal elite is not an option. It is the dictator's law. The entire memorizer system is based upon Orwellian "groupthink."

All dictatorships that have schools use the memorizer system. People are indoctrinated not to believe in themselves but rather to believe in someone else. It is a system that instills learned helplessness and dependency upon the authority of the elites. Here the shallow and pretentious get away with calling themselves "intellectuals." Here the abusers trick and force victims into harming themselves for a good grade. But elitist bullies are terrified that the public will discover its own ability to think freely. The elites know that free people change the world by changing themselves. Once the public has awakened from its trance, the criminal ruling class will lose its power to lie, cheat, and steal.

Under a statist-run education, by the time a student graduates, the natural ability for free thinking has been beaten out of him or her. Students are trained to deceive themselves. They are taught that it is their duty to submit to external authority every time. They believe what they are told to believe and, worse yet, they do what they are told to do without question. Whatever the establishment is, they blindly defend it. The memorizers are purposely trained not to outcompete the ruling class. They are trained to know their box and stay in it. The memorizers are taught that they are free...as long as they do what they are told.

Identity politics are used to divide, separate, and conquer students. This process is used to destroy each student's personal identity and reshape it into a centralized force. Gang mentality is the opposite of individualism. With a gang mentality, students become reactionaries. They are taught to react to the stimuli of external authority and not to think for themselves. Only external stimuli are allowed and praised.

Internal stimuli of oneself is mocked and outlawed by political correctness. Thinking for oneself is called "selfishness" and is forbidden. Individuals are categorized, labeled, and separated into so many false and superficial identities that a person can no longer be at one with him- or herself and others. Human beings are divided into different groups to destroy their sense of individualism and commonality with all.

As students advance, they are placed into specialized compartments, where they memorize and become aware of only specific areas of knowledge. When placed into groups, students are then taught to follow the confirmation bias of that gang. This further destroys each student's individualism. The memorizer system is compartmentalized so the elites can play the various gangs against one another.

The memorization system encourages unthinking "learning" procedures. It is a system of self-delusion based upon pseudoscience. Some of the most successful students of the brainwashing system become psychopaths. Putting psychopaths into powerful positions leads to disaster. They have no empathy for others. Not only do they destroy themselves, but they destroy those around them. Psychopaths are compulsive because they surrender to the imaginary. It is their own make believe that destroys them. The climate fraudsters are example A.

Brainwashing instills mental laziness and mindless habits. Students who graduate with honors from the memorization system have a hard time doing their own critical thinking. They get to secure positions in life only when they relearn how to think for themselves. They were trained to be employees. They were trained to be dependent upon someone else to provide jobs instead of creating their own jobs. A real education trains people to become employers. Instead, students are encouraged to suppress their own genius and creativity. They are trained to remember facts. Little cheap computers do that now. This is why college degrees are depreciating. It also explains why, in the real world, some top students end up working for average students.

At best, after graduation, memorizers are only capable of grasping an occasional small piece of reality. Those who fail to shift out of their reactionary education will never be able to see the entire picture of basic reality but are only able to glimpse small pieces of it. The graduates' indoctrination to external authority can be so complete that they cannot even recognize their own abusers. They are programmed to do what they are told to do whether it makes sense or not. After graduation, the continuous harangue of political correctness is used to keep the brainwashed in line. With this system, people are persuaded to believe in the unreal. This is the prime foundation for an Orwellian society. War is peace, and right is wrong.

A nation is no greater than the sum total of all its citizens. Each individual's self-worth and creativity totaled as a whole are what really matter, not the possessions of the ruling class. When the public's awareness increases, the criminal ruling class will cease to exploit and enslave. This is why the understanding of honest reality is so vital for a free society to survive.

The greatest truth about unity is that people are unified by individuality, which is all-inclusive. People who belong to a gang make themselves exclusive. They become gang members, special interests fighting for power over all others. At the top of that power structure are the criminal elites who profit off the gangs that fight one another. For them war is profit, and more war means more illicit power and more illicit money.

In a nutshell, the whole purpose of the memorizer education system is to dumb down the public so the ruling class can stay in power and exploit humanity. The rulers know that the best victims fight for their own victimization. They sell themselves on the fraud that is right in front of them, never knowing they were trained to be trusting fools. In state-run schools, the main lesson taught is submission to state authority.

The ruling class knows it cannot buffalo or bamboozle *everyone* with every Ponzi scheme, but for them that is OK. As long as the public cannot escape from all the anticivilization swindlers, the public remains under some control by the ruling class. Some people see through the money swindles and tell everyone they can about those crimes against humanity.

But some of these same people are trapped in the environmentalist science swindle.

Some are hooked so badly that they even participate in promoting the climate swindle. They are blind to the fact that they rightfully rail against the state-sanctioned counterfeiting of money, but at the same time they champion the state counterfeiting of science. Usurpers do not care which hoax the public falls for as long as the public prostrates itself to some Ponzi scheme. Those who believe in different idiocies are still fools who can be exploited by con artists. That is what holds the anticivilization together. But now, honest reality puts the pattern together for everyone to see the entire corruption of the anticivilization and how it works. As long as people are willing to accept tyranny, it does not matter what form of statist fraud they choose. The same elites will be their masters, and the criminal class will stay in power.

The initial purpose of the memorizer system is to keep the students in a childish state and so brain damaged that they cannot tell real from unreal. The system destroys their inner selves and their essence of being human. Students have to replace that void with an external authority. This keeps students looking for a parent or leader to tell them what to do. Under this system, people can become so brainwashed that they will not believe reality even when proof is handed to them. As an old con artists' joke says, "Are you going to believe me, or are you going to believe your lying eyes?"

All this is why environmentalists have hijacked their way into the school system. Knowing how difficult it can be to persuade or trick adult minds, the environmental con artists have gone after the immature and undeveloped minds of children. Now lurking in the nation's classrooms are environmentalist child molesters. Instead of molesting the defenseless little bodies of children, these con artists terrorize innocent little minds with lies and fear. In school, innocent children are taught that Mommy and Daddy's carbon footprints are killing all the polar bears. Innocent little children are told that their own carbon footprint is responsible for the glaciers melting, which will allegedly flood coastal cities and drown or displace millions of people. Little children are taught to hate their parents, hate themselves, and hate everyone around them because people are evil and destroy the planet with carbon dioxide. The environmentalists accuse the children and their families of horror after horror and insist that the only way to repent is to surrender their lives and property to the all-knowing environmentalist saviors. And it has worked...until now.

It does not take a supermind to defeat those who memorize and recite lies. Liars are not clever people. If they were truly smart, they would not lie in the first place. Liars give the public everything needed to expose and ruin the liar. One of the best signs of a memorizer is his or her continuous demand to obey authority instead of listening to reason. They keep trying to use political correctness to control speech. When speech is controlled, ideas are controlled. Those who devalue reality with false information also devalue society. Lies are a cancer that infect and corrupt everything they touch.

Even for the honest, it is easy to make a mistake once in a while. Bad information is all around us. But the professional liars willfully and continually deceive for a living. However, to prove someone is lying is easy for the intellectually honest to do. Just take the reverse of what someone is saying and hold it next to the original statement. One claim will be wrong and dishonest, and soon a pattern develops—a pattern of honesty or a pattern of lies. Professional liars are a lot more dishonest and wrong on important things than regular people are. Free people do their own research and keep asking honest questions and investigating the things that interest them. Free people save themselves.

Honest reality is discovered by switching from a memorizer system to a comparative educational system. Indoctrination freezes knowledge and can even destroy it. Making true comparisons requires honest testing of theory and honest debate, the opposite of memorizing somebody else's words. The debate is not settled, there is no false censorship, and new knowledge is pursued. Bad ideas and corruption become exposed and are thrown out, while good ideas are pursued. This is the opposite of the hocus-pocus climate swindle in which blind acceptance is the order of the day.

Memorizing is not education. Learning to be an honest critical thinker is a true education. Remembering things is important, but being able to honestly think is of greater importance.

Part Four

Honest Reality

This last section is a blend of summation and informative technique. The world's greatest orators have long known that it is best to repeat the important parts of a speech so that they would not be soon forgotten.

> ***No army can withstand the strength of an idea whose time has come.***
>
> **—Victor Hugo**

A lie is a bluff against reality. Liars have no real substance. Every lie bullies and abusers tell makes them weaker and weaker in the real world. Eventually the liars become so pathetic that no one trusts them or their worn-out lies. The irony of the uncivilized is that they degrade themselves with every evil act they commit. It may not seem that way at first and, in fact, might even look like the evil will go on forever. But that is not the case. All Ponzi schemes eventually collapse because of their own corruption.

Exploiters of mankind are at the mercy of their own fraud. Their own lies are weapons that can be turned and used to destroy them. Professional liars unknowingly arm their intended victims with suicide buttons that control the liars' own survivability. The criminal class is really comprised of weaklings. Those who live by the lie are destroyed by that lie. Honest reality is an intellectual technology that cannot be stopped by anyone. As more working people learn how to control criminals with the criminals' own ignorance and deception, life for the civilized gets better and better. All Ponzi schemes can be eviscerated with honest reality.

The anticivilization is a society of mass mental illness. It is where fraud and superstition are called "science" and "justice." It is where those who tell the truth and expose fraud are called "evil" by the corrupt. But every day, more and more people see through the swindlers' games. The criminal ruling class's occupation of free society is falling apart.

The most powerful court of all is the court of public opinion. No corrupt government court system can withstand a full frontal assault from the court of public opinion. The government's courts *must* bow to the honest reality of the public or be replaced. A revolution of morality is a revolution the anticivilization cannot possibly win. The uncivilized cannot compete with a truly civilized society.

The anticivilization cannot be fixed. It has to be eliminated and replaced with real civilization. The criminal class is irrelevant, unneeded, and most of all, unwanted. The cleansing process starts with exposure of the climate swindle. Honest reality is the nullification of fraud. People are waking up to the fact that master parasites offer only violence and failure.

Environmentalist leaders do have a purpose in life—they serve as the bad example. They are deniers of reality, and they are haters of humanity. Climate fraud is a giant mind control stunt set up to destroy free society. The scheme is designed to enslave and denigrate human life by controlling all energy supplies. These usurpers have made billions of dollars off of their climate hoax, and now their time is ending. The only way they could maintain their fraud would be through censorship and the threat of violence.

Right now is the best time to be alive. The greatest and most powerful of all revolutions is happening now. This is the time of accelerating honest reality. This is where the individual rises to full power and true ability.

Centralization and its brutality are starting a death spiral. Centralizers have put all their eggs in one basket, and that basket is tipping over. Piece by piece, group by group, the criminal class is being exposed and broken up.

Honest reality is the individual's only real form of self-protection. It empowers the individual, and this frightens the ruling criminal class. As the truth is learned, the parasites lose the hosts they feed upon and starve to death. Honest reality is the antidote to the anticivilization. Honest reality is the Rosetta Stone of life that empowers humanity. In the real world, the con artists ultimately are proven to be no greater beings than those they tried to rule. To destroy evil, it is necessary to understand what it is and how it operates. Meanwhile, the criminal class will keep trying to convince the world their criminality does not exist.

There is a terrible secret the criminal leaders desperately try to keep hidden from the public. The secret is that all those who live in the anticivilization are guaranteed to fail. From top to bottom, the uncivilized are weak, dull-witted, corrupt, and in a suicide pact with evil. There are no great leaders in the anticivilization. They are all bullies, bluffers, and disgusting losers. To eliminate these evildoers, simply give them time to fulfill their destinies of destroying themselves. But that is much too costly in innocent life, time, and treasure. A much better way to eliminate the criminal class is to push their self-destruct buttons instead of waiting for them to do themselves in. The self-destruct button that liars are completely frightened of is called *honest reality.*

The evolution of mankind's thinking upsets the ruling class's best-laid plans to dominate forever. The status quo is like a small heard of carnivorous dinosaurs too fat to run and too slow to think properly. Now the über- predators are the prey. In the civilized world, criminals and their corruption are a food source and a source of public entertainment. With the exposure and destruction of each Ponzi scheme, society gets stronger, healthier, and wealthier. Truth takes away everything the liar has stolen.

The criminal class cannot exist when people stop obeying their orders. Con artists cannot exist when people stop believing their lies. When the master parasites are proven to have no real moral authority, their reign of stealing and terror is over. This is why they try to control the flow of information and education.

Anyone who understands how the memorizer system works can own its graduates. Like Pavlovian dogs, they are controlled by external authority. Whoever learns which external authority is in control of the student can ring that student's bell, flipping his or her switch on and off at will. The uncivilized choose their own masters of make believe, and make believe is easy to hijack. The ruling class is fully aware of this, and it is how it stays in power. The status quo plays the unknowing against one another to keep them confused and weak. Fantasy is the source of the master parasites' power over others, and they do not want anyone to learn that their magic powers are total fraud. The leaders of the climate hoax are the perfect example of fraudsters trying to control everyone else with a trance of mass hysteria. In the real world, there is no magic—there is only cause and effect.

Memorizers and the abused are trained to get their marching orders from "politically correct" authorities, carrying out those orders regardless of reason or reality. They might argue opinion, but they do not argue real facts. This is a fatal flaw. They gave up their own willpower to think for themselves. They were trained to obey someone else's orders.

People who work for a living can usually figure out what is real faster than those in academia. Academics led the Nazis and their eugenic policies. Academics led communists and their social justice thieving. And now academics lead the environmental climate swindle with pseudoscience supporting their fraud.

Most people in the world have a strong sense of humanity. They just want to be left alone so they can build, create, and enjoy life. Only a vicious minority of the population thinks they can live by lying, cheating, and stealing. Criminals in the world are greatly outnumbered by good people. And most important, the civilized are exceptional to the criminal class.

The truth is always directly in front of the public. The problem is that people are taught to ignore reality. But that is changing. Fewer and fewer people worship blind obedience. More and more are thinking for themselves instead of repeating con artists' lies. People are escaping from the anticivilization and are discovering the real world, the civilized world.

Peaceful change is needed. Once the criminals are publicly exposed, they will have their day in the courts. There are different courts and different legal actions to make the criminal class accountable for their crimes. The only thing a con artist has to work with is their credibility, and that is a value of pure illusion. When the con artists' worthlessness is publicly exposed, their swindle is over.

The anticivilization is a perpetual war machine. It is at constant war with the individual because the individual human being is the real source of power. If the uncivilized are allowed to usurp more power, the rest of society suffers and is weakened until it eventually collapses. The uncivilized are in a total war against every free individual on earth. Every chance they get they will try to destroy individual rights. For the criminal class, war against the individual is profit. War against reality is how the usurpers enrich and empower themselves at the expense of the good and innocent. But in the real world, this is all reversed—right side up.

To break the individuals' will power and to control them, usurpers instill feelings of powerlessness and hopelessness over victims. Individual aspirations are secondary, while the collective is everything. People are made to feel that they are a grain of sand on a vast beach, helpless against the smashing waves. But in reality, one grain of sand can jam the illusionary gears of the ruling class. One clear, lone voice can instantly become a majority view. Individualism is kryptonite to the anticivilization.

The anticivilization is a society of delusion, bluffers who desperately hope no one will call them out and expose their malevolent frauds. The environmentalists, like all other con artists, are cowards who try to bluff their way through life and get something for nothing. They have no power but the power they steal from others. When exposed, these cheats will do what they do best—tell more lies. And when those lies are exposed as fraud, they tell *more* lies. That is all they know how to do.

Every Ponzi scheme rots from the inside out. Each swindle is destroyed by its own corruption, because make believe cannot sustain itself. In the real world, liars weaken themselves by giving up their dignity and replacing it with provable fraud. They destroy their own credibility. Liars seldom use real facts, because the facts are not on their side. Instead they use innuendos to slant people's views away from reality and toward the unreal. All the con artists can do is try to bluff and intimidate. They are terrorized of being called out for their treachery, because just about everything they do is fake. Those who obfuscate reality are destroyed by what they tried to hide.

In the final analysis, con artists are their own worst enemies. They give their victims everything needed to destroy the con artist. Liars build cases against themselves. Every lie they tell is an indictment against them. But connivers still try to tell any lie that will make a quick profit.

It is relatively easy to break down any Ponzi scheme. They can be taken apart point by point because so many lies are begging to be exposed. It takes a lot of energy to keep a lie alive, and environmentalists have told so many big lies that they are spent. That is one of the great ironies—eventually criminals suck themselves dry as their own fraud consumes them.

Most of the basic information needed to collapse and destroy every Ponzi swindle is already in the public domain. Each pillar the scheme is built upon must be honestly reexamined and analyzed. The results can be shocking, the corruption profound. That is the design of all swindles. They are totally fake and worthless, just like the liars who sell them.

"Manmade global warming" is a deliberate fraud. It is a gigantic swindle of provable lies. Real scientists do not use censorship to win debates. To the unknowing, the words of the con artist may seem to be solid statements, but when held up to the truth, they are recognized as nonsense. The best way to transcend these lies is to put the focus on honest reality. The anticivilization is a world of continuous conflation, the secret of all Ponzi schemes. It is the backbone and framework of all swindles. It is where the unreal is claimed to be real and the real becomes fake. For example, fraud-based manmade climate change becomes as real as natural climate change and is conflated as the real thing. The con artists automatically call their fraud "climate change," and according to the same liars, nature's greenest life-giving gas becomes a toxic poison. But in the real world, conflation is exposed as fraud. Lies cannot be supported by truth. Lies can only be made to look real with more lies. Every Ponzi is a stack of criminal lies.

Here are some of the secret pieces of *honest reality* environmentalist leaders are terrified the public will discover and break the trance of fraud:

- The current volume of ice at both poles is now increasing.
- Banning DDT killed millions of innocents in the third world.
- Carbon dioxide is not poison or a toxic pollution.
- There are natural snowstorms in the tropics and there are glaciers by the equator.
- The "global greenhouse theory" is a 150-year-old superstition.

- In the last decades the polar bear population has exploded fivefold.
- The last glacial system melted thousands of years ago, and earth is at an interglacial time period.
- Floating sea ice that melts does not change the water level.
- There is no trapped heat in the atmosphere. The only place trapped atmospheric heat can be found is in a bag of fairy dust.
- Even the image of smokestacks supposedly belching carbon dioxide into the atmosphere is intellectually corrupt.

Carbon dioxide is colorless and odorless. What the environmentalists' pictures show is mostly water vapor. Trying to stop earth's natural carbon dioxide cycle is as insane as trying to stop earth's water cycle. Either action would destroy all life on planet earth.

Meanwhile, the most damaging piece of reality the environmentalist leaders are desperate to hide is the fact that they are in the game for personal illicit profit. Eventually these swindlers will be held accountable for everything they have stolen and destroyed.

Science is never settled. There is always something new to discover and learn. The manmade global warming hoax is not for the protection of humanity. It is exploitation by counterfeit "saviors."

In the anticivilization, policy is based on delusions. Just like politics and economics, the field of science is currently dominated by massive dishonesty. Well over half of so-called scientific fact is based on fraud and cannot be replicated. It is done to protect corrupt careers and keep the dishonest collecting illicit paychecks. This is done at the expense of the health and safety of the public so a handful of elites can enrich themselves.

Honest people do not fear the truth, but the dishonest are terrified by it. Honest reality destroys the pretend world. The climate swindle is a pseudo problem being solved with pseudo answers. The uncivilized have no idea of what they are talking about. They speak in tongues of gibberish and talk about "carbon footprints." There is no sense in arguing with the willfully stupid. The best way to deal with deliberate fools is to use their own foolishness against them. Reason defeats emotion by using emotion against itself.

Like all swindlers, environmentalists have no real evidence to support their lies. Conjecture is not reality. These charlatans sell imaginary crimes against nature. They are so backward that even their most basic arguments are Orwellian "doublethink."

Honest reality cannot be unlearned. Those who read about it cannot deny it and pretend that it does not exist. Honest reality ends the monopoly of the elites. Honest law destroys liar laws. It is common knowledge that any contract entered into that is based upon fraud is an illegal contract.

Fraud breaks contracts. Illegal contracts are null and void in a civilized society. Honest law says they do not have legitimate authority and are not to be obeyed. Liar laws are illegal laws by which one party deliberately defrauds another. This means the anticivilization is powerless in the real world.

There is something more powerful than the hardware of war—the software of war. Honest reality is the essence of survival. Those who don't have it perish. Only the honest can own reality. Honest information is technology that destroys liars and tyrants. This is why the ruling class is desperate to control free speech. It is also why, throughout history, every time the pseudo academics get into power, they purge the real intellectuals from society, just as the climate fraudsters are trying to do now.

The only thing that keeps liars alive is their ability to bluff about reality. They can ignore reality all they want, but their denial will not change the truth. The usurpers' hardware can be dangerous, but it can also be destroyed. The software is the crucial part of the package. As long as the software, the fatally flawed bad idea, is allowed to exist, the simpleminded who believe in it will keep reconstituting themselves and come back to do more harm. Without the software of the big lie, there is no reason for the liars' existence. Thanks to honest reality, con artists no longer have a reason to exist. Their ploys will become extinct.

Deep down, the ruling class is fearful of the public. These fakers know that the average working person is just as smart, if not smarter, than they are. This is why all usurpers attempt to control and censor as much information as possible. The truth will destroy them. People succeed when they do their own research and ask honest questions. When people do their own thinking, Ponzi schemes fall apart. Honest reality is discovered when individuals do their own due diligence. Trust, but first verify what is real.

Honest reality is not a centralized power. It is an individual power of free-thinking people. It is the golden key that demystifies fraud. Those who really want to discover and use honest reality can do so on their own. There are only a couple of basics to follow:

- Do not take political claims or the so-called truth at face value. Con artists depend on this. Instead, research and think about those claims. Consider the opposite of what is claimed to be true and then decide which opinion is honest.
- The truth is the opposite of what a liar says it is. Look for contradictions, falsehoods, and superstitions. Soon a pattern will emerge and it will become evident whether or not the messenger and the message is based on truth. Regular people who think honestly protect themselves with the most powerful force in the universe—honest reality. Reality is the one thing liars will never have on their side, and it guarantees their failure.

Everyone from the anticivilization is a pretender. They are not what they claim to be. Con artists are empty bags, just like the schemes they are selling. They may have the appearance of invincible, unstoppable power and all its trappings, but in the end they are all bluff. They are fatally broken counterfeits. The last thing a con artist wants is a trial in a court of honest law. Environmentalists have already lost their trial in the court of public opinion. Most people already know that the climate hoaxers are liars and thieves.

No matter what the phony saviors say is their reason for existing, it is always about money and power. Money is the representation of human labor. Next to health, money is the control of everything. The lazy and evil want to get money by not working for it. The world is in a constant battle between those who create real values that enable their neighbors versus those who create fraud to harm and disable their neighbors. The climate fraudsters have their own cult of evil that sticks together because those without morals are easy to bribe.

The antiscience climate scam is over. Now the world knows how the trick works, and its magic is gone. It was all a worthless joke told by thieves who are exactly what they claim not to be. They are the ones who declared war on nature. They want to outlaw nature's greenest gas. Not only do lower carbon dioxide levels starve plants of food, but if it were to drop below 180 ppm, plants would die off, taking animals and humans with them. Environmentalism is an antiscience cult devoid of common sense.

Environmentalist con artists have nowhere to hide. All of their lies, demands, and extortions are publicly recorded. Their crimes against their neighbors are now visible for everyone to see. An unstoppable retribution will come for these criminals.

The real dangers to these scam artists are all the people who discover that they were robbed and swindled. The believers and victims, who were used and abused, will demand justice and reparations from the climate crooks. Those people will potentially go after the criminals and sue for damages. The government and public will claw back the money they were defrauded out of. Skeptics have long known climate swindlers were liars and they refused to be involved with them.

In a civilized society, there is no sanctuary for the criminal. Those who rule by terror live in fear of their own aggressions coming back to them. They will never know for certain how retribution will visit them. All they can do is worry about the certainty of their coming payback. They are prisoners to their own corruption. Those who live by the lie eventually suffer the same fate as that lie.

For those who want to opt out of the anticivilization, it can be done: Stop believing the liars and their lies, and get as far away from the Ponzi schemes as possible.

The best way for people to empower themselves is to take control of their own labor. Free people vote every day with their own actions and money. They just go ahead and do creative things. They don't wait for permission to

live. They deal only with true value-givers who improve life with honest goods, services, and new information. And they help to expose those who try to exploit the innocent. Silence about evil is acceptance of evil.

The summation of the anticivilization is easy—ignorance is strength, witchcraft is science, and personal responsibility is for peasants. People trapped in the anticivilization do not have to worry about being placed on someone's list. Everyone is already on the predator's menu. There is no such thing as being safe in fantasy land.

Honest reality is the death of tyranny and centralized power. Great innovations and evolutions are not started by group thinkers. True creators refuse to wait for the majority to catch up. Once the good and innocent wake up to reality and stay focused, they surprise even themselves as to what they can accomplish.

Real knowledge has been suppressed for so long that mankind will now start to have breathtaking leaps in knowledge and in the understanding of how things really work. Human evolution will surge because honest reality is obliterating more and more evil every day.

Of course there are real pollution problems that need to be addressed. For example, mercury and sulfur dioxide are real toxins that need to be monitored and contained. But the con artists have contaminated honest science. The climate fraudsters have willfully set back real scientists and their work for years, but this is all changing for the better.

Honest reality creates mass epiphanies. For instance, those who put pure carbon dioxide into the air should be congratulated for making earth greener and more livable for all plants, animals, and humans.

When people realize their potential to do good things, they will do them. A new level of honest awareness transmutes the corrupt and evil. The anticivilization is trembling and shaking apart, one Ponzi scheme at a time.

All Ponzi schemes are made up of odious debt. Every single demand from con artists is nothing but fake debt and false obligation. Every so-called claim owed to the criminal class is in fact pure fraud. No one owes criminal liars anything except a quick trial and a trip to prison.

Honest reality is the greatest false debt eliminator in the world. For too long, swindlers have gotten away with enslaving victims with counterfeit debts and obligations. These scammers run around telling their intended victims "you owe me," or "you owe it to yourself," or "you owe it to others." Then these brazen thieves demand money. One favorite environmentalist debt scam was their ridiculous carbon footprint swindle. These fools didn't even know that carbon and carbon dioxide are two different things. Their own conflation rules these pompous clowns.

Carbon dioxide is two thirds oxygen and occurs naturally through geological and biological actions. All animals, including fish and insects, emit carbon dioxide. Even soil releases it.

Everything that dies and decomposes produces carbon dioxide. Lakes, rivers, and oceans dissolve limestone and release carbon dioxide into the air. The carbon dioxide cycle is as natural and vital to life as the water cycle. Earth, by itself, is a carbon dioxide factory.

In the real world, even basic math proves the climate schemers are dead wrong. The official amount of carbon dioxide in the air is currently about four hundred ppm. That means that out of one million molecules in the air, four hundred of them are carbon dioxide. Since nature puts 96 percent of all carbon dioxide into the air, man is responsible for sixteen ppm out of one million total units. Sixteen divided into one million equals 62,500. Therefore, according to those who side with manmade global warming, one molecule of nature's greenest gas out of 62,500 molecules is causing global warming. Environmentalist leaders are barking-mad lunatics.

It gets worse. The climate crackpots claim that by shaving off part of that one tiny trace molecule they can save the earth from disastrous climate change. They promise that when their extortion money is paid to them, they can control earth's thermostat. These people are beyond ridiculous. They are dangerous.

Also contrary to environmentalist doctrine, carbon dioxide does not trap and hold heat forever any more than a frying pan holds heat forever. Heat passes through both. The less molecular density there is, the faster heat passes through. In the real world, carbon dioxide is not flammable and is used to put out fires. Also, carbon dioxide gas can be used for cooling and for freezing. Anyone can prove it.

All that is needed is: a carbon dioxide charged canister (like those used in air guns), a release mechanism (like a bicycle tire inflator), and a thermometer. The experiment only takes a few seconds to demonstrate the cooling ability of carbon dioxide. For a dramatic visual effect, spray a couple short bursts of carbon dioxide gas on the thermometer. The temperature reading will go from room temperature to below freezing very quickly. (All experiments are the sole responsibility of those doing the experiment. These are not toys).

Also, when carbon dioxide is frozen (at minus 109.3 degrees Fahrenheit), it becomes a solid called dry ice. Shippers and grocers use dry ice to keep food products frozen. Dry ice is so cold that it will burn bare skin when it is touched. Do not play with it.

The Coup D'état

There is only one way to deal with those who refuse to be reasonable. Hold them to the same standards they demand from everyone else. Call their bluff. Tell them "you first." The environmentalists have to ignore the reality that during the Jurassic period, carbon dioxide levels were over two thousand ppm and earth was covered in lush green vegetation and there were record amounts of animals both large and small.

Those who believe carbon dioxide is a dangerous toxic pollution that destroys the earth when it goes over 350 ppm can only do one thing to prove to the world they are right. They must hold their breath. Every word that comes out of their mouths is carried on a carpet of over forty thousand ppm of carbon dioxide. The carbon dioxide they exhale with each breath is over one hundred times their own proclaimed point of no return. Environmentalists who refuse to stop exhaling carbon dioxide admit to the rest of the world that they are criminal frauds. According to the environmentalists' own dogma, they are killing mother earth with each breath they exhale. (Do *not* hold your breath).

Earth's climate has been—and always will be—variable. Trying to reduce man's miniscule introduction of carbon dioxide into the atmosphere to stop climate from changing is an obscene epitome of pseudoscience.

The awakening of the public scares the criminal elites. These cheaters know that their own weapon system of lies can be turned and used against them. The destruction of the manmade global warming cult is the perfect template to use against all other Ponzi schemes. In the real world, the anticivilization is rotten, corrupt, and beyond repair. The master parasites' tricks of using fear and then promising to lead victims to safety for a fee do not work as well as they used to.

One of the cheaters' favorite tricks is to use conflation against their victims. Thanks to a defective education system, too many people still believe that correlation automatically equals causation. It does not. For example, as the temperature increases, the number of chirps a cricket makes also increases, and therefore, because of correlation, when a cricket chirps more, the temperature will increase. Claiming that correlation always equals causation is very sloppy thinking. It is just as foolish as saying that because carbon dioxide is associated with heat, it thereby causes heat. In actuality, the reverse is true. Carbon dioxide is the result of heat breaking down carbon. In the real world, the fields of science, economics, and politics are wide open to anyone who is intellectually honest and willing to explore reality. All a person needs to do is escape the inverted thinking of the willfully ignorant.

Mankind's overall effect on the temperature of the earth is so slight that it cannot be accurately measured. Manmade global warming is the biggest antiscience extortion racket in history. It is a worldwide criminal enterprise designed to bankrupt and destroy the civilized world for the pleasure and profit of elitist criminals. The environmentalists have stolen billions of dollars using their steaming stack of lies and liar laws. They have already committed mass genocide by indirectly killing millions of human beings, and now they demand more power and money.

Liar laws are the building blocks of tyranny and political correctness is the glue that holds the dictatorship together. Wherever there is political correctness, there is corruption, and the truth is being hidden from public view. Not only does reality shatter the pretend world, it is the antidote for political correctness.

Truth destroys all illusions. Once people see the real truth, they cannot unlearn it. Honest reality is superior to all lies because the truth lasts forever and lies are illusions that fall apart over time. The manmade global warming theory is a house of cards built out of deliberate fraud and willful stupidity. But the climate con artists' free lunch is now over. They can no longer pretend to be pious saviors. From now on, whenever an environmentalist starts to bray about the dangers of carbon dioxide and its global warming effects, they will have to explain the indefensible and the impossible.

In summary, environmentalist make believe can not withstand reality. Winter is caused by shortening the total amount of sunlight and summer is caused by increasing sunshine. The sun is the main controller of earths' surface temperature and climate.

Honest reality is the technology of awareness, and it cannot be stopped. Genuine knowledge makes the criminal elite extinct. Honest reality reverses the liars' lies. In an honest world based on truth, it is the environmentalists themselves who are the real deniers and flat earth worshipers.

The End

Genius is nothing more than the ability to recognize the obvious. When they apply themselves, most people can experience genius.

David Noonan

Bibliography

"American Physicist Joins Attack on Global Warming Theory." Greenie Watch. June 29, 2010.

"An American Coup D'état—First World Nation Style." Cognitive Dissonance. November 9, 2010. ZeroHedge.com.

Ball, Tim. "Global Temperature and Data Distortions Continue." Canada Free Press. July 29, 2010.

Benbrook, Bradley A. "Climategate's Next Phase: False Claims Act Lawsuit." Andrew Breitbart. January 2010.

Booker, Christopher. "The Fiddling with Temperature Data Is the Biggest Science Scandal Ever." February 7, 2015. Telegraph.co.uk.

Caruba, Alan. "Big Green Lies Are Imploding." Climate Realists. August 8, 2010.

"Climate Science Is in the Crapper." Real Science. May 22, 2013.

"CO2 Is the Magic Gas That Makes Plants Grow." Jonova Global Warming. April 14, 2010.

"December 31 Global Sea Ice Area Was the Largest Ever Recorded." Real Science. January 2, 2014.

Delingpole, James. "US Physics Professor: 'Global Warming Is the Greatest and Most Successful Pseudoscientific Fraud I Have Seen in My Long Life.'" October 9, 2010. Telegraph.co.uk.

Durden, Tyler via Gordon T. Lang. "An Orwellian America." March 15, 2013. ZeroHedge.com.

Fenig, Ethel C. "If You Have Bedbugs, Thank Al Gore." American Thinker. September 9, 2010.

Gass, James. "Obamacore: The Nationalization of K-12 Education." Breitbart. January 27, 2014.

Gosselin, P. "Climate Scandals: List of 94 Climate-gates." Notrickszone. Fall 2010.

— "Gate Blowup! Come on in Gate Lovers!" Notrickszone. May 20, 2010.

— "Giant of Geology/Glaciology, Christian Schluchter Refutes CO2…Feature Interview Throws Climate Science into Disarray." Notrickszone. June 9, 2014.

Goddard, Steven. "Global Cooling Notes from Reader 'SUYTS.'" Real Science. November 11, 2010.

— "One Reason Why NOAA Is Lying about US Temperatures." Real Science. January 12, 2013.

"Latest Peer-Research: Solar/Cosmic Ray Combo Is Driving Climate Change, Not Human CO2 Emissions." C3 Headlines. 2010.

Markson, Lisa. "Rachel Carson's Ecological Genocide." July 31, 2003. Frontpagemagazine.com.

"Memo to Senators: US Climate Scientists Are Manipulating Temperature Data, and You Also." C3 Headlines. 2010.

"Mercury Dumping." The Daily Bayonet. Climate Change Fraud. January 1, 2010.

Michaels, Patrick J. "Earth Day 40." World Climate Report. April 21, 2010.

Milloy, Steven. "DDT: A Weapon of Mass Survival." May 4, 2006. FoxNews.com.

Moore, Steve. "Settled Science Scam." March 11, 2015. Forbes.com.

Morano, Marc. "Execute Skeptics! Shock Call to Action." Climate Depot. June 3, 2009.

— "Left-Wing Environment Scientist Bails Out of Global Warming Movement. Declares It a 'Corrupt Social Phenomenon…Strictly an Imaginary Problem of the First World Middleclass.'" Climate Depot. July 26, 2010.

— "New York Times Publishes Cartoon about Killing 'Deniers' with Icicles." Climate Depot. February 23, 2014.

— "Special Report: More than 1000 International Scientists Dissent over Manmade Global Warming Claims." Climate Depot. December 8, 2010.

— "The Big Lie: Skeptics Funded by Big Oil—No, the Alarmists Are." Climate Depot. July 24, 2013.

"New Zealand Climate Scientists Admit to Faking Temperatures: The Actual Temps Show Little Warming over Last Fifty Years." C3 Headlines. 2010.

Nova, Joanne. "Climate Money." SPPI Original Paper. July 21, 2009.

Olson, Joe. "It's Clouds Illusions That I Recall." Climate Realists. June 30, 2010.

O'Sullivan, John. "NASA in Shock New Controversy: Two Global Warming Reasons Why." Climate Realists. May 27, 2010.

– "Top Climate Scientists Speak Out on the Satellite-Gate Scandal." Canada Free Press. August 16, 2010.

Page, Lewis. "New Science Upsets Calculations on Sea Level Rise, Climate Change." November 28, 2012. TheRegister.co.uk.

Roberts, Donald and Jasson Urbach. "DDT Causes Reduced Cognitive Ability...in Journalists." February 27, 2014. Forbes.com.

Seymour, Julia A. "And That's the Way It Was: In 1972 Cronkite Warned of 'New Ice Age.'" March 5, 2015. Newsbusters.org.

– "Exploding Children in Eco-Group's Video Fails to Upset Liberal News Media." Business & Media Institute. October 6, 2010.

Shaw, Jazz. "Is A Lot of Scientific Research Just...Crap?" January 19, 2014. HotAir.com.

Sheppard, Noel. "Breaking: UN IPCC Official Admits 'We Redistribute World's Wealth by Climate Policy.'" November 18, 2010. Newsbusters.org.

"Special Report: A-Z Climate Reality Check." Climate Depot. December 7, 2011.

Taylor, James. "Global Warming Alarmists Caught Doctoring '97 Percent Consensus' Claims." May 30, 2013. Forbes.com

Unruh, Bob. "Confirmed: Global Warming Is 'Settled'—as a Scam." Worldnetdaily. April 20, 2010

Voisin, Ronald D. "Please, Let's Wake Up to the Most Important AGW Facts." Climate Realists. April 16, 2010.

Washington, George. "Americans Are Finally Learning about False Flag Terror." November 24, 2013. ZeroHedge.com.

Watson, Paul Joseph. "Global Warming Alarmist Call for Eco-Gulags to Re-Educate Climate Deniers." September 20, 2010. PrisonPlanet.com.

Watts, Anthony. "About That Overwhelming 97–98 Percent Number of Scientists That Say There Is a Climate Consensus." July 18, 2012. wattsupwiththat.com.

"Welcome to the Insane Asylum—Making Reality Fresh Daily—Chapter 4." Cognitive Dissonance. Zero Hedge. June 6, 2010.

Additional Reading:

Red Hot Lies
By
Christopher C. Horner
Regnery Publishing, Inc.

The Deniers
By
Lawrence Soloman
Richard Vigilante Books

The Really Inconvenient Truths
By
Iain Murray
Regnery Publishing, Inc.

Made in United States
North Haven, CT
04 November 2026

21768316R00046